SPOTLIGHT ON NATURE

HUMMINGBIRD

PAMELA DELL

CREATIVE EDUCATION · CREATIVE PAPERBACKS

Published by Creative Education and Creative Paperbacks
P.O. Box 227, Mankato, Minnesota 56002
Creative Education and Creative Paperbacks
are imprints of The Creative Company
www.thecreativecompany.us

Design by Blue Design, Inc.
Art direction by Tom Morgan
Edited by Grace Beltowski

Photographs by Alamy Stock Photo/Scott T. Smith / Danita Delimont, Agent, 18; Dreamstime/Dewald Reiners, 10, Judy Kennamer, 16, Rinus Baak, 17; flickr/Biodiversity Heritage Library, 11; Getty Images/Alex Saberi, 29, Arthur Morris, 23, BirdImages, 6, Dorit Bar-Zakay, 21, George D. Lepp, 4–5, Hal Beral, 12, James F. / 500px, 29, ©Juan Carlos Vindas, 14, 29, Judy Unger, 15, LordRunar, 3, 6, 29, VladSt, 8, 10, 14, 16, 20, 22; iStock/JohnPitcher, 27; Pexels/Pixabay, cover, 1; Shutterstock/Ondrej Prosicky, 9; Unsplash/Izumi, 24; Wikimedia Commons/Andy Morffew, 21, Paul Danese, 28

Library of Congress Cataloging-in-Publication Data
Names: Dell, Pamela author
Title: Hummingbird / by Pamela Dell.
Description: Mankato, Minnesota : Creative Education, Creative Paperbacks, [2026] | Series: Spotlight on nature | Includes bibliographical references and index. | Audience term: juvenile | Audience: Ages 10-13 Creative Education and Creative Paperbacks | Audience: Grades 4-6 Creative Education and Creative Paperbacks | Summary: "Zip into the hummingbird's world with this nature title for middle-grade wildlife lovers. Informational text pairs with a narrative about a single bird family to spotlight the wild animal's life cycle, supported by infographics and a timeline of developmental milestones"— Provided by publisher.
Identifiers: LCCN 2024048395 (print) | LCCN 2024048396 (ebook) | ISBN 9798889896227 library binding | ISBN 9781682777886 paperback | ISBN 9798889897026 ebook
Subjects: LCSH: Hummingbirds—Juvenile literature
Classification: LCC QL696.A558 D45 2026 (print) | LCC QL696.A558 (ebook) | DDC 598.7/64—dc23/eng/20241213
LC record available at https://lccn.loc.gov/2024048395
LC ebook record available at https://lccn.loc.gov/2024048396

Printed in India

CONTENTS

MEET THE FAMILY

ANDEAN HILLSTARS of Peru

High in Peru's Andes Mountains lies a central plateau of **puna** grasslands and rocky ground. Few plants grow here. The climate at this altitude is dry, cold, and windy. Snowcapped peaks rise above pastures and rocky, open ground where herbs, shrubs, and lichens grow. Frost and drought are challenging for all life up here above the tree line. Still, many animals make their home in these grasslands. The Andean cat, the Andean fox, and the puma prowl this area. The large Darwin's rhea and many other birds live here, too. One of these is the small Andean hillstar hummingbird.

A female Andean hillstar defends her territory alone. She has built her nest alone. The nest is secured to the face of a rock hidden by bushes. Inside this cup-shaped nest, she has **incubated** her two jelly bean-sized eggs for 20 days, keeping them safe and warm. But now change is coming. Very soon, her chicks will break free.

CLOSE-UP

Eyes

Compared to other birds, hummingbird eyes take up more space on their heads. Placed on the sides of the head, their eyes give them nearly 360-degree vision. Their vision is so sharp that they see an incredible range of brilliant colors unknown to humans.

CHAPTER ONE

LIFE BEGINS

Hummingbirds are among the smallest birds on the planet. In fact, the Cuban bee hummingbird is the tiniest creature in the whole **avian** world. There are about 360 hummingbird **species**. While there are slight differences among them, they share many of the same traits.

All hummingbirds live only in the Western Hemisphere. They exist from Canada in North America down to Chile and Argentina in South America, and they are found everywhere in between. They also live in the Caribbean Islands. Many North American species **migrate** south when the weather grows cold.

Most hummingbirds measure only 3 to 5 inches (7.6–12.7 centimeters) long. Females are generally about 15 to 20 percent larger than males. They need to be bigger so their eggs can develop inside their bodies.

Hummingbird beaks, also called bills, are slender, needle-like, and usually very long compared to those of most other birds. This helps them

ANDEAN HILLSTAR HUMMINGBIRD MILESTONES

DAY 1

- First egg hatches
- Length: 1 inch (2.5 cm)
- Eyes are shut
- Pink or gray featherless skin

DAY 2

- Second egg hatches

dip into deep, tube-shaped flowers to get at the nectar they prefer to eat. Many red flowers are a rich source of nectar. This makes them especially attractive to hummingbirds. Some hummingbirds eat as much as 12 times their weight in nectar every day. They also eat tiny insects such as spiders and aphids. But, as with most wild animals, predators stalk hummingbirds, too.

Nestlings are at the greatest risk. They are often prey to snakes and bigger birds. If they make it through their first year, their chances of survival improve. But cats and hawks then become major predators. The praying mantis insect is also a vicious enemy. With its sudden, lightning-quick strike, hummingbird death is almost certain.

Hummingbirds rarely flock together. They ferociously defend their own chosen territories. When it comes to nest building and raising chicks,

CLOSE-UP

Tongue

Hummingbirds sip flower nectar with long, grooved, W-shaped tongues. Tiny hairs on the tip of the tongue help them lap up their meal at a rapid rate of up to 18 times per second. When not in use, the tongue is coiled up.

FEATURED FAMILY

Welcome to the World

The mother Andean hillstar has built a small but bulky nest no bigger than a walnut shell. Inside are two eggs, each weighing no more than a paper clip. One day, a small hole appears in one of the eggs. Little by little, the hole expands into a crack. With a little more work, the **hatchling** finally escapes. A day or so later, the second hatchling works its way out into the world, too. The babies weigh less than a dime. They are weak and blind. Their skin is grayish-pink, and they are nearly featherless. But life has begun.

females do all the work. They pick their nest locations carefully. Mother hummingbirds eat many times daily. They also feed their chicks often throughout the day. So they must build their nests near food sources. Different hummingbird species have their own nest-building techniques. But most have a base made of stretchy spider web strands. This helps the nest expand as the chicks grow. Until they're ready to fly, hatchlings are entirely dependent on their mother for food and warmth.

DAY 9

- Eyes open

DAY 10

- Pin-like feathers begin to grow

CLOSE-UP

Bill

The hummingbird's bill is built long for dipping deep into flowers. It is usually curved, but this varies with species. The lower jaw can widen at its base and bend downwards. When battling each other, male hummingbirds sometimes use their bills as stabbing weapons.

FEATURED FAMILY

First Meal

The two baby hummingbirds have hatched. Their mother has cleaned the nest of all broken eggshells. Now it is time to eat. The hatchlings need to be fed about every 20 minutes. First on the menu is protein, or meat, to build strong bones and bills. Their mother makes many trips to deliver small spiders and bugs. She gradually starts feeding them sugary flower nectar for energy. To feed them, she sticks her long bill into their mouths. She brings up, or regurgitates, her swallowed food directly into their stomachs. Dinner is served.

Hummingbirds are most

ATTRACTED

to bright-colored flowers, especially red or orange.

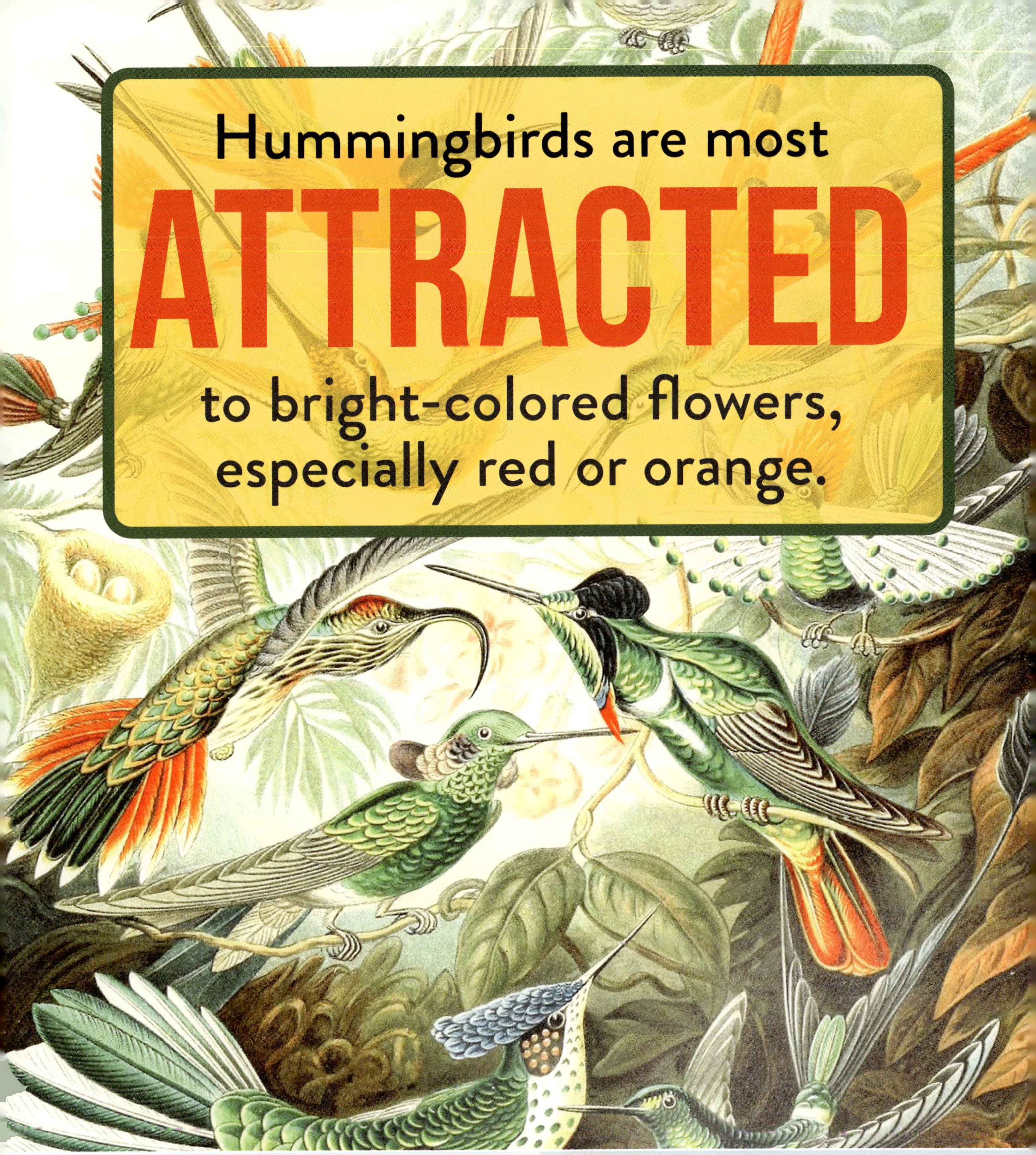

DAY (12)

- Mother begins leaving chicks alone, even on chilly nights

DAY (14)

- Bill has grown much longer
- Real feathers begin to grow in

CLOSE-UP

Feathers

Most hummingbirds have ten long primary feathers on each wing. They have six secondary feathers on each side. Most species also have tails with ten feathers. Unlike all other birds, hummingbirds have no down feathers to trap heat and warm them.

CHAPTER TWO

EARLY ADVENTURES

About two weeks after hatching, hummingbird chicks begin to exercise their wings. They are preparing for their first flights. In another two weeks they will be entirely independent. Except during their mating period, they will not need others of their kind. In fact, small though they are, hummingbirds have an aggressive instinct. If there are plenty of feeding spots in a garden or at a hummingbird feeder, a few might settle there together. But usually they will fight off the competition.

Hummingbirds are also territorial by nature, especially males. If they find a good patch of flowers to feed on, they won't want to share. Every flower holds only a small amount of nectar. It takes some a whole day to make more. And no flower lasts very long itself. But defending their food sources may be easier for hummingbirds than trying to find new ones. If another hummingbird comes near, trouble is likely.

DAYS 14–16

- Starts vigorously exercising wings in preparation for first flight

CLOSE-UP

Feet

Hummingbird feet, or claws, have three forward-facing toes and one backward-facing toe. The inner surfaces are ridged, giving them a good grip on flower stems and feeders. Hummingbirds can't walk or hop like other birds. They shuffle sideways to move along their perch.

FEATURED FAMILY

More to Learn

The **fledgling** stage comes early for hummingbirds. At just three weeks, the two tiny birds now have all their feathers. Their tail feathers have not fully grown in, but they are nevertheless able to fly. Still, the fledglings remain close to their nest. They are not done learning yet. For the next week or so, their mother will guide them in how to survive. She will show them where to find nectar-filled flowers. She also teaches them how to catch insects while in flight. This is an important skill, called "hawking."

Hummingbirds are not only fierce. They are sometimes surprisingly brave. When one of them becomes aggressive, even large hawk moths and bumblebees need to watch out. Hummingbirds have been known to occasionally go after hawks and other birds much larger than they are. Even small mammals might be subject to attack if they get too close to a hummingbird's food source. The hummingbird will try to scare off the intruder however it can. The little bird might zoom at the enemy, chattering and squeaking. It might spread its tail feathers and dive at the target. Hummingbirds will do anything to win.

DAYS 20–25

- Fully fledged, or able to fly
- All feathers grown in
- Leaves nest without mother prompting

CLOSE-UP

Gorget

The bright, colorful patch of feathers on a male hummingbird's throat is called the **gorget** (*GOR-jit*). Researchers believe these flashy feathers help the bird distinguish itself among species. It also aids in attracting mates. Female hummingbirds rarely have a gorget.

FEATURED FAMILY

Give It a Try

High in the central Andes Mountains, one of the Andean hillstar hummingbirds has only recently fledged. Unlike other hummingbirds, his mother perches on the plants she feeds on. Perching saves her energy in the cold mountain air. Her chick copies her behavior. He's still a youngster, so he hasn't yet developed the brilliant green gorget that will identify him later. But he has a built-in ability to withstand the harsh, high-altitude climate. And now his mother has led him to an orange-flowered shrub, her favorite. It's a delicious discovery, and now his favorite taste, too.

When it comes to defending a food source,

HUMMINGBIRDS RISE TO THE BATTLE.

(1) MONTH

- No longer fed by mother
- Now considered an independent adult
- Migratory species ready for first migration

CLOSE-UP

Body fat

When the weather turns cold, some non-migrating hummingbirds put on weight to stay alive. As they eat, the sugar they consume changes into fat. Hummingbirds that lack enough body fat or **plumage** go into a sleeplike state called torpor, similar to **hibernation**.

CHAPTER THREE

LIFE LESSONS

The average lifespan of most hummingbirds is about three to five years. Some species live much longer. But many hummingbirds live in forests and other remote areas. This makes it difficult to know how long they live. Some studies estimate that more than half of all hummingbirds die within their first year. Predators, including domestic cats, are a constant threat. For those that migrate, the long flight back and forth can be exhausting. Flying into windows often kills hummingbirds as well. Yet one of their greatest causes of death comes from backyard feeders. Thousands die each year from spoiled homemade sugar water and dirty feeders. Toxic commercial sugar water, or "nectar" containing red dye, chemicals, and preservatives, can kill hummingbirds, too.

If they make it through year one, hummingbirds have a better chance of living a longer life. At one year old, they are sexually mature and ready to mate. As with many bird species, male hummingbirds look much flashier than females. Females need to lie low because they have eggs and babies

DAYS 30–40

- Fledging occurs in species in colder climates

DAYS 45–60

- Mother birds of some tropical species stop feeding their young to protect. But the males' brilliant, **iridescent** greens, purples, and blues catch the females' eyes.

Male courtship rituals are elaborate. The birds' specialty is their stunning and action-packed aerial displays. They get better at these performances over time. When an interested female appears, the show begins. It may start with the male hovering, singing, and "dancing." Some species twirl rapidly or perform fancy tail moves in midair. But the really spectacular feat is the bird's "courtship dive." The male may fly as high as 130 feet (39.6 meters) up in the air. Then, with wings beating about 200 times per second, he plunges straight down at 55 miles (88.5 kilometers) per hour or more. If the female isn't impressed, she flies off.

FEATURED FAMILY

This Is How It's Done

One of the newly fledged hummingbirds is hungry and ready to go. She's about 5 inches (12.7 cm) long. She weighs about 0.28 ounce (7.9 grams). Every day she works hard to get enough to eat. She may stop to drink flower nectar hundreds of times a day. She feeds mostly on flowers from shrubs and low-growing cactus plants, and sometimes on eucalyptus. She's learned other ways to snag dinner, too. Insects crawling on plants are tasty treats. And she hawks them right out of the air.

CLOSE-UP

Flight

Unlike other birds, hummingbirds don't flap their wings. Instead, their wings rotate. This allows them to fly backwards and upside down. Using a unique figure-eight stroke, they can also hover. During some species' mating displays, air vibrating strongly in their tail feathers can cause loud "pop-like" sounds.

1 YEAR

- Most die within this time
- Ready to mate and reproduce
- Males perform elaborate mating displays
- Females select mates and build nests

Hummingbirds usually lay two eggs, which take 11 to 18 days to hatch. All hummingbirds learn to fly and find food from just a few weeks old. They also have large brains and a remarkable memory. They remember exactly where to find flowers they have visited before. They know how long it takes a flower to fill up again with nectar. Sometimes they even remember human faces. Humans in return are always looking out for these beautiful creatures.

CLOSE-UP

Humming

The hummingbird's humming sound is unique among all animals. It is caused by the rapid upstrokes and downstrokes of its wings. It might warn enemies away or impress a potential mate. It's the sound that gives these birds their name.

FEATURED FAMILY

Practice Makes Perfect

The other young hummingbird is now ready to face anything in his search for food. His super-high **metabolism** means he must drink from as many flowers as possible. The sugar in nectar gives him his energy. He flits from one plant to the next, perching just long enough to get his fill. He has claimed his territory and is a tough defender of it. When a large bumblebee flies near, the hillstar chases it away at top speed. Other creatures are not welcome in his feeding ground!

Male HUMMINGBIRDS go all out in their flashy mating dances to impress watching females.

3–5 YEARS

- End of life for Allen's hummingbird and other species

7–11 YEARS

- End of life for Anna's, ruby-throated, rufous, and other species

CHAPTER FOUR

HELPING HUMMINGBIRDS SURVIVE

Early Spanish explorers in the New World had never seen hummingbirds before. They called them ***joyas voladoras***, or "flying jewels." With their brilliant, iridescent coloring, many hummingbird species have magical names to match: Crimson topaz. Red-tailed comet. Rainbow-bearded thornbill. Purple-crowned fairy. Even in countries where they don't exist, hummingbirds have many fans. These little birds are a gift to the world that needs to be protected.

Of all the amazing species, almost half live in the "Equatorial belt." This is an area of the globe that goes from 10 degrees north of the Equator to 10 degrees south of it. Colombia has the most hummingbird species with more than 160. Ecuador has 130 species but the greatest number of hummingbirds overall. Fewer than 25 species live in the United States. Chile and Canada each have fewer than 10 species. And hummingbird populations are in decline. Some species have already been lost, such as the Brace's emerald of the Bahamas. And there are many threats besides predators to those that still exist.

One major threat is habitat destruction. Land where hummingbirds live is being cleared of forests for timber, farming, and human housing. People illegally capture and kill hummingbirds to use in "love potions" or as good luck charms. They have also been used in some human medicines. **Pesticides** kill insects that hummingbirds need for food. And climate change affects flowering plants that give hummingbirds nectar. Some plants may bloom too soon or too late—or not at all—disrupting the bird's natural feeding cycle.

Many organizations and private citizens are working to protect these precious birds. It is important work. The American Bird Conservancy and the International Hummingbird Society are two organizations putting much effort into helping hummingbirds. Both are engaged in projects aimed at protecting some of the most endangered species. They work with governments to protect habitats. They also have programs that educate the public.

Hummingbird research at many universities is ongoing. Only relatively recently, researchers began understanding the complicated science of how hummingbirds fly. Another area of research is the remarkable hummingbird brain. Scientists study how the brain understands and reacts to the world. Still other studies focus on hummingbird migration patterns and feeding habits. All this research will help people learn how to protect these beautiful flying jewels.

FAMILY ALBUM

SNAPSHOTS

The **calliope hummingbird** is the smallest breeding bird in the U.S. and Canada. Females sometimes steal other birds' nesting materials to build their own.

The **Anna's hummingbird** has the fastest, most dangerous courtship dive of all. They drop about 385 body lengths per second, or faster than the Space Shuttle entering Earth's atmosphere.

When temperatures drop, most **ruby-throated hummingbirds** leave Canada or the U.S. They fly nonstop across the entire Gulf of Mexico to southern Mexico.

The endangered **Juan Fernandez firecrown** is only found on Robinson Crusoe Island. This tiny island lies 400 miles (644 km) off the coast of Chile.

The **long-tailed sylph**'s bill is shorter than most. It sometimes pierces a flower's base to get at the nectar.

The tail of the **marvelous spatuletail** is unlike any other. It has just four wire-like feathers. Two of these end in flat blue-violet discs called "spatules."

In 1951, the only known specimen of the **glittering starfrontlet** was found. Another one of this very rare species wasn't discovered until 2004.

Ecuador's **black-breasted puffleg** is one of the most endangered hummingbird species. As of 2019, only 250 to 999 adult birds were known to exist.

The **rufous hummingbird** migrates annually between Alaska and northwest Canada to Mexico—a remarkable 4,000 miles (6,437 km) each way.

The **blue-bearded helmetcrest** is one of the world's rarest hummingbird species. Found at the highest altitudes of Colombia, it is also extremely endangered.

The **Santa Marta sabrewing**, a hummingbird of Colombia, hadn't been seen since 1946. But in 2010, after more than 60 years, a researcher found one.

WORDS to Know

avian — having to do with birds

fledgling — a young bird that has just learned to fly

hatchling — an animal that has just recently broken out of its egg

hibernation — a long period of deep sleep experienced by some animals in cold weather, when food is scarce

incubate — (of a bird) sit on eggs, keeping them warm as the chicks inside develop

iridescent — displaying glowing colors that seem to change when viewed from different angles

metabolism — the chemical processes within the body that keep animals alive

migrate — move from one region or territory to another according to the season

nestling — a baby bird not old enough to fly or leave the nest

pesticide — a poison used on plants to protect them from being destroyed by insects

plumage — the entire feathery covering of a bird

puna — a high grassland plateau in the Peruvian Andes Mountains

species — a group of living things that have shared characteristics and are able to reproduce with one another

LEARN MORE

Books

Groc, Isabelle. *A Hummingbird on My Balcony*. Victoria, B.C.: Orca Book Publishers, 2025.

Riggs, Kate. *Hummingbirds*. Mankato, Minn.: Creative Education and Creative Paperbacks, 2022.

Rose, Rachel. *Hummingbird*. Minneapolis: Bearport Publishing Company, 2024.

Websites

"For Kids!—Hummingbirds." Audubon Delta. https://delta.audubon.org/conservation/kids-hummingbirds.

"Hummingbird." Britannica Kids. https://kids.britannica.com/kids/article/hummingbird/353273.

"Hummingbird Spot for Kids." Hummingbird Spot. https://www.hummingbirdspot.com/h-spot-for-kids

Documentaries

Carroll, Don, and Noriko Carroll, dirs. *First Flight: A Mother Hummingbird's Story*. Concept Images, 2009.

Ferraro, Carlo, dir. *Hummingbirds of Costa Rica*. Ferraro Nature Films, 2016.

Reddish, Paul, dir. *Hummingbirds: Jewelled Messengers*. Free Spirit Films, 2012.

Note: Every effort has been made to ensure that any websites listed above were active at the time of publication. However, because of the nature of the Internet, it is impossible to guarantee that these sites will remain active indefinitely or that their contents will not be altered.

Visit

BANDELIER NATIONAL MONUMENT

At this nesting sanctuary, you'll find calliope, rufous, black-chinned, and broad-tailed hummingbirds.

15 Entrance Road
Los Alamos, NM 87544

LAKE HOPE STATE PARK

In the summer, park officials demonstrate how to hand-feed ruby-throated hummingbirds.

27331 State Route 278
McArthur, OH 45651

TOHO CHUL PARK

Visit the hummingbird garden to watch Anna's and Costa's hummingbirds sipping nectar.

7366 North Paseo del Norte
Tucson, AZ 85704

UC SANTA CRUZ, ARBORETUM AND BOTANIC GARDEN

Trek the Hummingbird Trail or visit during Hummingbird Month in March.

1156 High Street
Santa Cruz, CA 95064

INDEX